Re Definition of Insanity

By Adam Ristow

Introduction

Do you have any idea, what the human mind is capable of? What a human can do to another human? Humans are delusional, most are weak, and don't obtain the aptness to confront mental wounds, or they endeavor too anyway. If you're reading this, you have discerned the illusion of civilization.

There's more than meets the eye to humans, and everything in life that we produce or that has been created. People with the highest i.q's are rather isolated from their peers, or they isolate themselves from their peers, and or society, they often have so called "mental disorders", they lack social skills or solicitude, but where there is a paucity of empathy, intelligence is pervaded. No such thing as a surplus of education.

This book is designed to reveal the psychology, and physics of civilization, and our life paths, and the science behind it all, you will be blown away. Civilization wasn't a mistake, it was the greed that possessed us that was the mistake. It doesn't help with our current government either.

Do you wonder if humans can be trusted without control, and order? Can there be a balance? Are all humans insane? Who are the real "crazies" of the world? The craziest of people such as Sigmund Freud, Albert Einstein, Sir Isaac Newton, Galileo, Boor, and so on, these people were all labeled crazy, and delusional in their time. When in reality, they were henceforth of their time, and it took the world a century for society to verge upon, and see that these people were right.

There has been a phlegmatic and subtle movement over the past couple of years. People, civilians, are starting to rise throughout the country. You have seen it on CNN, daily news, in the media, several controversial movies are starting to trend. Rioting has been occurring in several states through the vicinity of the country.

Johnathan Lee, who resides in the U.S., is a genius in his mid-20s, and doesn't obtain a college degree, yet Johnathan has mental abilities that society labels him as mentally ill. These mental illnesses allow him to discern what society does not. Johnathan Lee is a mental, and trance medium. He has the ability to telepathically receive messages from his spirit guide Sarah, he obtains abilities to hear, and feel entities from the spiritual side. The messages are emitted telepathically, and will arrive in his "mind's eye" in form of a word, voice, or image, anywhere from a few seconds, to a few minutes in the future.

Hence the word mental in mental medium, a medium is higher than a psychic. Johnathan Lee has dreams foretelling the future about month in advance. Mediums relay messages from the spiritual world to our living world. The trance part of his abilities is when his spirit guide will literally seize his conscious mind, and talk for him, utilizing his voice, but talking for him, and using particular vocabulary he doesn't use, the other part of his mind will be aware of this happening.

Johnathan inherited these abilities from his relatives before him. His relatives aren't mediums, but rather possess attributes of a psychic. I must first tell you this, Johnathan developed his abilities at the age of 20, and he is now in his mid-20s. His abilities were dormant all his life, until a certain event happened when he was 19 that opened the door to his abilities, and allowed him to see the illusion of society.

His mediumship abilities acquiesces his mind to discern the illusion of society. Johnathan often feels social phobia in public, and sometimes in his own family social conventions. Johnathan has isolated himself from society, and yet he loves to have fun, but only with selective people. He feels he is too smart to let himself live in the illusion everyone else does, or if they do see beyond the illusion, he wonders why they choose to continue to live in the illusion.

Johnathan is a functioning psychopath according to psychology; he has been mentally wounded, but is self-aware. His father deserted him twice, at 5, then again at 23, the 2nd time hurt worse; he was old enough to know what happened. Johnathan's father chooses alcohol and Jonathan's step mom over Johnathan. I will not dive into his life story, that is a whole other book unto itself, as so is everyone's.

Johnathan's mother is a perfectionist, and has severe control issues with her children. Her maternal instincts are high; she is either over critical or over smothering. She does not possess the psychological education to take care of her own issues, let alone comfort Johnathan. Not entirely her fault though, by time Johnathan Lee reached the latency stage, his mother had been severely mentally wounded as well.

Johnathan obtains attachment issues due to his father deserting him at 5 years old, and multifarious factors, variables, and his life, led him to narcissism, obsession, voyeurism, misogynistic, sadistic {in terms of his womanizing} and twisted sexual fantasies, and is a functioning alcoholic. Johnathan has psychotic tendencies, but is consciously a great man, he covets helping people. His psychotic tendencies are more so with women, he does not lay a hand on them, just womanizes them. These are his so called mental illnesses in a nutshell, which Johnathan does not believe in mental illnesses.

Now, through history the so called "crazies" were later proved to be of the highest of intelligence. The ultimate question is "am I insane, or are others?" Johnathan's answer is, we all are. As I mentioned earlier, Johnathan does not obtain a college degree, he has an outstanding memory, and ability to retain education. His abilities to derive a conclusion, and has created mind-bending theories in theoretical physics.

Johnathan Lee has read all of Freud, Einstein, and Newton. Johnathan has always loved, and studied physics, and psychology, well; psychology is the science of the mind. Johnathan would sit in his room at 20 years old, and just write for hours, about his day was, his observations, and he realized Freud was right, and so was Einstein. Freud is idol for psychology.

Johnathan's father once told him Einstein's definition of insanity, which is doing the same thing over and over expecting a different result. That was a factor in what inspired him to look deeper inside what insanity is. As Johnathan grew older, he observed more, and would write what he observed, and felt. He did his best when he simply wrote on paper, and found many similarities between the universe and humans. I will walk you through the experience, and how Johnathan grew in his mediumship, and how his mediumship allowed his mind to see, and do the extraordinary. Then I will take you through his new definition of insanity.

During the spring time when he was 19, when he was sleeping in his room in the early to mid-morning, {yes he is a night owl} he felt footsteps walking into his bedroom at a certain time every morning. He felt the vibrations of the footsteps, then the footsteps stopped at his bed, and he felt the entity leaning against his bed, and watching him, anxiety struck Johnathan. At first Johnathan thought it was his mother who was walking in his room to wake him up, so he asked his mother if she was walking in his room, and she said no.

Now at this time, Johnathan didn't realize the entity was his late brother in law who died several months ago. This occurrence kept reoccurring every single morning to the point where Johnathan decided to sleep downstairs. Johnathan decided to conduct an evp in his room during the evening. During this evp, Johnathan asked "who are you?" "What do you want"? "Give me a sign you are here with me." "Speak into the device I am holding in my hands".

After about 10 minutes of the evp, Johnathan called it quits. Then listened back, and what he heard sent shivers through him. He listened to the evp in his room in absence of electromagnetic radiation, and sound, and at 2 minutes Johnathan heard "its jah" now, Johnathan thinks the entity didn't complete the sentence. He strongly believed it was his late brother in law. At 5 minutes, he heard "Johnathan", which was his name of course.

That was what rendered him speech less; Johnathan froze with shock, and excitement. He had his mother listen to the evp as well, she had the same reaction. Later the next day his older brother arrived to drop off his son to spend the night. Johnathan had his brother listen to his evp, his brother reacted the same way Johnathan did. His brother said after the evp "it knows your name"!

Johnathan's older brother, and Johnathan auscultated to it in Jonathans room in silence, once up here after the evp, the two brothers were leaving the bedroom, but Johnathan's brother received a message in his cranium that didn't want him to leave, he quickly turned around at the frame of the bedroom door, and said "it doesn't want me to leave". That alarmed Johnathan, he wasn't sure if it was really their late brother in law now.

Johnathan's brother and Johnathan stood in the center of his bedroom, and felt a circle of cold air circulate around them. Their hairs on their arms erected, cutis amnesia, and shivers, seized Johnathan, and his brother said "if this is our brother in law, I forgive you for everything that happened between you, and my sister, its water under the bridge, you are at peace now. Johnathan didn't say anything to their brother in law. The brothers were misty eyed after that experience.

It was after that experience Johnathan didn't hear from his late brother in law for a long time. Johnathan, and his brother in law were very close, he took Johnathan to baseball games, haunted houses, and movies. It was a bitter sweet experience for Johnathan, and his brother.

It was this experience he believes that unlocked the door to his mediumship, and along with what I mentioned earlier about his relatives on his mother's side obtain certain psychic abilities too. At 20, Johnathan started hearing indistinct voices in his head, and aloud, and he could feel the emotions of others, and entities. Johnathan came to his mother, and his sister for questions he had. They told revealed to him what this was, it was only a matter of when before his abilities unleashed.

The family is a league of psychics with 1 medium. Johnathan's mediumship abilities didn't unleash until about another year at 21. At 20, another occurrence he will always remember, was when he was in college for the first time, he parked his car in a desolate spot in the parking lot. As he was walking out of the campus one afternoon, he walked to his car, far away from the rest of the cars, and Johnathan opened his car door, threw his back pack in, and sat inside.

Soon as he sat in, he heard in a whisper clear as day "can you hear me"? Johnathan immediately turned around, thinking there was someone next to him outside, and there wasn't anyone outside. He checked his cell phone to make sure it wasn't on, and it wasn't. He then knew it was a ghost trying to communicate with him.

As time went on, Johnathan would ask, and apprise his mother, and sister about occurrences that happened. Biggest factor that Johnathan didn't like was telling the entities to go to "the light", when Johnathan does not believe in "the light". So his mother told him about a psychic that she has seen before, and certain training methods she has used, so Johnathan used these techniques as well, they are as follows. Johnathan would go into his room, turn off all distractions, and sit on his bed, relax his skeletal system, all muscles, tendons, and take big, long, deep breaths, inhale, and exhale, and his mind's eye, he would picture a funnel over his head, and apprise all entities to come to him, that he is inviting them in, to communicate with him, talk to him, tell him what they covet.

The funnel was to symbolize the voices going down the funnel, and into his temporal lobe. Johnathan would practice this about 30 minutes at a time, now, during this time he never received a response while in his practicing mode, the voices always came after. Johnathan knew if he were to tell people about his abilities, people would peg him as crazy, and insane. Many times in conversation, he would be able to pin point how a person felt, and foretell what the person would say a few seconds ahead of time, he thought he was foretelling the future.

In conversation at 22 years old, Johnathan used his ability on a girl, on a date, he asked his date when her birthday was, and his date told him to guess. Johnathan did not guess, but used his ability, he would excavate his mind, and the first response that hit him was May 17th ,he already knew the year, she was 19 at this time. Johnathan's date's qualitative characteristics could be characterized as mesmerized, and she said "how did you know that?" Johnathan replied "it's all up here", pointing to his head, smiling. About a month, or so later Johnathan found out he doesn't obtain the ability to read minds, this is his spirit guide communicating with him, therefore classifying him as a medium.

About a month, or so after his date, he had a dream, in this dream, he resided in his apartment at this time, sitting on his couch, and he turned his head to the left, and saw a female mannequin. The female mannequin turned her head to the right to meet Johnathan, waived her hand, and said "I have been watching you". Johnathan heard this voice in his dream, when he woke up, he apprised his mother, and sister about it as well. They told him to communicate what that dream was, later on he communicated, and the response wasn't what was expected. His response was this, Emma, his spirit guide who lived during the civil war.

When Johnathan received his response telepathically in his mind's eye, the name Emma came

in big, sparkling yellow letters.

This meant a spirit guide, the color of the letters in his mind's eye is in correlation to the

aura of the entity that the message is for, or came from. Now, Johnathan does not believe in

god, heaven, hell, angels, satan, or at least not in the way that we are all told. Johnathan has

formed his own beliefs about the spiritual world, and religion, and how one can believe in the

paranormal without believing in god.

At 20 years old, as I mentioned earlier Johnathan would sit in his room for hours, and

write about his daily thoughts, observations, and solving the equation, and basically answering

his own questions he had. Johnathan noticed that humans have many similarities with the

universe, and particles. At 20, Johnathan's father told him that Einstein's definition of insanity

is doing the same thing over, and over expecting a different result.

This intrigued Johnathan, and from there sparked his passion for psychology, he always

had a passion for it; he lived psychology, and believes it's the science of the mind, and

psychiatry is a crock. Pills do not solve anything, hence why the patient has to keep getting

refills. Johnathan worked on a theory about insanity for about 3 years off and on. He came up

with a new definition for it, by experimenting on himself by staying up for 1, 2, 3 days at a time

to study the effects of sleep deprivation, and how he believes there's no such thing. All his

studying, his own theories he derived up, and his experimenting on himself, he came up with

the new definition of insanity, in his own words.

Chapter 1

Levels and Types of Insanity

What is insanity? Albert Einstein defined it as doing the same thing over expecting a different result. Johnathan will go deeper, and define it as our wake state of mind. When in this wake state of mind, we are all on in our everyday lives, most of us have the inability to see the damage we inflict upon ourselves, or others from the choices we make. This state of mind is not the state of mind most think it is.

Most of us are born sane, the many instincts we have. Through the history of time, and as civilization began, and grew we have been environmentally conditioned to ignore our instincts. What happens when we ignore what we feel? The emotion manifests, that is how we have all become insane, along with the damage we inflict upon each other.

When we are in our wake state of mind, we are insane, when we are sleep deprived, or in the altered state of mind, or even asleep, one is sane.   Type 1 insanity is the wake state of mind, where we are all insane without any major traumas, through environmental conditioning to ignore our instincts, the individual has become insane, and the insanity manifested.   Type 2 is in the wake state of mind, and is when one has been mentally wounded. In this type of insanity is where the levels come in, which are mild, medium, and high.

In the altered state of mind when intensely "sleep deprived", or the "twilight" state, means in between wake, and asleep. It becomes more interesting the longer you're awake. The reason why, in this state of mind happens in the middle of asleep, and awake is because of the multi universe theory, and Johnathan's theory on how we are creations of the universe theory. We, as in humans, are in the middle of the multi universes according to recent talk in the science field. More often than not in this state of mind, we can access memory better, and our cognitive functioning is better, our problem solving is higher, the percentage used by our brains is much higher due to beta, alpha, and delta waves that emerge during sleep, and our emotions are not as high. Time is nonexistent when in this state of mind, when one has been awake for over 24 hours, 8 hours will have gone by in just 5 minutes it feels like.

The images or sounds that inflict insanity type 2 are accepted by the appropriate part of your brain, through a series of electrical signals of axons, and dendrites to a nucleotide in your D.N.A, and literally altering a part of you. Johnathan will now explain type 2 insanity, and the levels with it. You have probably observed in conversation that people often will not admit they're wrong, or blame their choice or the action on others around them. This is a sign of insanity.

In psychology, this is projection, as humans, we have guilt. Every emotion is derived from the subconscious, from the amygdala. The subconscious is located in the lower region of the brain. When a person blames another person for their own choice or action, that person is projecting what they feel.

That person that feels the guilt, and is pinning the blame on another person. The inability to see this is a sign of mild insanity. Small actions and choices are a sign of mild insanity. It's human instinct to covet social interaction, and to desire the subconscious closeness with another individual. Humans are instinctively insecure. We have to be taught introspectiveness to be secure within our minds.

Next up is medium insanity; this level has to do with more personality disorders, or people living in a delusion, such as believing there is a god, and living in a denial state. Women are prone to this level of insanity even without any mental wounds. Women covet to be told what they want to hear, even though they know the truth, and many other women are more attractive than them.

Women are prone to live in a delusion, women know men think other women are more attractive, but they desire to be told that they are the most attractive, or the most wonderful person. Another factor is primitive, many women think certain male celebrities are their counter parts, and become easily jealous other women. The more women handle stressful situations, and are hurt men, etc. they become misandrists.

This is not sexist, this is biology, and psychology, and from what history has shown. All these factors root to estrogen, the female hormone. The word estrogen derives from Greek, and the figurative meaning is sexual passion, and or desire. Estrogen is produced by tissues, and organs such as the liver, adrenal glands, or breasts.

People with medium insanity have been wounded, and due to environmental conditioning their I.D. has chosen certain methods to cope with the anxiety they once felt for the wound. Such as misogyny, which is the objectification of women due to mother issues a son has, or misandry, which is the objectification of men due to father issues a daughter has.

Obsession for a human is another coping mechanism. However, obsession can be, or is an un-healthy behavior, depending on the actions you take for your obsession for a person. Obsession can lead to a delusion, or living in denial about the person you are obsessed with. Usually the person with obsession has a neglectful parent, or two neglectful parents, and a method to cope with the wound he/she covets the closeness of the opposite sex as a coping mechanism for the neglectful parent, and is obsessed with a male friend for the absent father, or mother. The relationship between parents, and their offspring are of high significance.

Voyeurism is another coping mechanism, which is the arousal of spying on people who are engaging in coitus, or naked, without their knowledge of the person spying. Now, this can go in couple different directions, perhaps just the instinctual desire to see the opposite gender naked, or engaging in coitus, or perhaps the person spying has developed absurd sexual fantasies, and is experiencing difficulty acting on those fantasies. Are you observing that some of these "disorders" are sexual, or what made the mental wound was sexual? Which just proves Johnathan's whole point, people's I.D.'s chose these methods for a coping mechanism to deal with the anxiety they once felt for the mental wound that inflicted it. It is all instincts, humans want to act out from our instincts, but it is not socially acceptable in society.

As Johnathan mentioned earlier about how women feel emotion derived from the amygdala stronger, due to this, many women who have been wounded by men, or in their life paths, and experienced difficulty to confront their wounds, therefore their minds chose to live in a delusion by believing in god. Now this can happen just as well to men, men can become misogynists, and women can become misandrists.

Johnathan will not discuss all personality disorders, they are all just labels designed to categorize us. Some are more disturbing than others, but the disorders are just coping or defense mechanisms inflicted by mental wounds, which were inflicted by someone else. Humans are weak, and delusional, and many struggle to confront their wounds, this is all in our wake state.

High insanity is among those who kill, rape, molest others for a hobby, and don't descry they inflict. Serial killers or pedophiles that engage in coitus with children, that is high insanity. Johnathan feels very strongly against pedophiles, possible due to his mental wound of being molested as a child by another fellow school mate. Pedophilia is the sexual arousal for children. A 19 year old, and a 16 year old is not pedophilia.

Johnathan has often asked himself who the real psychopaths are. The answer is there is, there is no real psychopath. We are all insane, like he said earlier. The conformists display an image, present themselves to appear ok, when they are not.

Killing people is wrong. Only in the most extreme situations is it justifiable to commit it. The high insanity is when one is committing these terrible acts and does not descry. Not just killing, also raping, anything in the severe category. Human trafficking is another terrible act to commit. This is a coping mechanism for something very severe that happened in that person's life. So many theories about how this comes to be, but one answer that he can assure is, no such thing as the criminal mind, if pushed, we all have the ability to become high insane.

The levels of insanity are in our everyday life, you will know it when you see it. The level of behavior or choice's, sets the level of insanity they are. This is all in our wake state of mind, and was type 2 with the levels. People who have not experienced a mental wound are type 1 without any levels, but maybe mild insanity. Those people are still delusional, the answer lies ahead in a further chapter.

Even medium insanity reacts off instincts, depending on the act the person did. We can repress our instincts all we want, but our many instincts will always be in our D.N.A. We have both positive, and negative instincts. Maternal, Paternal instincts, tranquility, and peacefulness.

Can there be a balance? Can humans meet a mental balance of not too much of instincts, but not too much repression or environmental conditioning? Ever notice how people with so-called mental illnesses are more often than not highly intelligent, just need psychoanalytic therapy is all, which is due to civilization fearing intelligence, and creativity.

People who observe everything that is backwards, or feel socially phobic for various reasons are sometimes people with so called mental disorders, have a much higher cognitive abilities, and problem solving abilities, and lack in empathy or social skills as said earlier. The people who made the criteria for these disorders, are insane themselves. Humans often see what we feel. Hence those who created the criteria for these disorders, had many themselves, but failed to recognize it.

Psychiatry is not efficient. One big factor in why is that people who are under pills, have to keep getting refills. The solution has to come from the mind, years of mental training and therapy, only if the patient wants to be treated. Talking therapy is the best, not medication.

Chapter 2

Instincts, the justice system, and controversy

Now, with this chapter Johnathan has discerned that the government produced laws based on human's instincts, and he discusses many controversial issues in our country. This is all Johnathan's theory, so allow him to explain this one. Sigmund Freud the founding father of psychology wrote "The Ego and the Id" in 1923, in that book he describes how humans have an I.D. and an Ego. The terms Ego, and Super Ego are not how we use them today, the term Ego really means our conscious.

The Catholic Church, many moons ago decided to label our instincts as the 7 deadly sins. That we must pray to God to give us strength to be pure of the evil. You can't repress your instincts.

All emotions, and instincts derive from the amygdala, and the subconscious. In civilization we have been taught to find a different outlet that is socially acceptable to let out the anger or hostile feeling. This has much religious background to it, and has made a cultural impact on America over the past centuries.

The Superego filters out which impulse goes where, through a series of axons, dendrites, and electrical signals, and the Ego, during this process we feel anxiety. We want to act on the impulse, but due to laws, and punishment, and other factors, we choose the alternative, the environmentally conditioned part of us makes the socially acceptable choice to not act on the impulse. We all have our ways on dealing with the impulse.

One positive method to cope with the impulse is the reality principal. Use an image or a sound derived from your subconscious. From external stimuli, and use the image or sound, and picture yourself doing what the image our sound is doing. Also, understand that depending on what image our sound you're using, may be dangerous to use in reality.

Another part of our instincts are the paternal and maternal. Now, this is generally referred to about parents of offspring, but it works this direction as well. It is biology that men are the providers and women are the nesters or nurturers. Men protect the woman, and his family.

No such thing as a pervert, it is only our instincts. Our instincts are not socially accepted in the social norm. If you think to yourself yes there are perverts, Johnathan will say there are pedophiles. Pedophilia is not instinctual, that is insanity. That's a gift or a trauma gone horrible, and untreated. Humans are perverts.

Humans make dirty jokes, some more than others. That is one of our instincts, lust, slipping through the cracks of our I.D. and our Super Ego. Humans project from our I.D. See, our instincts don't originally exist in the external world, only in our internal world, internal meaning our mind. We repress our instincts such as lust so much we make so called dirty jokes as a coping mechanism.

All laws are designed to repress our instincts. John will go even deeper here and talk incest. Yes incest, believe it or not, there are many people who covet to engage in coitus with their relatives.   Humans are environmentally conditioned not to do so. On this earth where laws do not exist, exists human trafficking in own families.

Johnathan said this earlier, do you have any idea what the human mind is capable of? What a human can do to another human. Sadism and Masochism, these are coping mechanisms. A sadist is one who is sexually aroused by giving others physical or emotional pain. Masochists are those who are sexually aroused by receiving emotional or physical pain.

These are so called mental or sexual disorders. No, they are not, they're coping mechanisms. People that are sexual sadists and masochists, most likely have been physically or mentally abused in their childhood, and they never dealt with the trauma.

Sex itself is not dangerous. Thus what humans do with sex, or how we act on the desire.   Guns, smoke, alcohol, weapons, cars, money, you name it. None of these possessions are dangerous by themselves, humans are the danger.

Gluttony is an instinct, just the over indulgence of food or material, no such thing. We need to eat to survive. There are people who will envy those people who are gluttonous. Envy is an instinct along with the others that are the 7 deadly sins. Envy is just jealousy.

Now to avoid jealousy, as Johnathan mentioned earlier, jealousy implies insecurity, to be secure means examining yourself, everything you have accomplished positive and negative. If you're content with whom you are, and what you are, and your life, than nothing else matters. These are all the 7 deadly sins, which were formed by the Catholic Church. Some of civilizations background has to do with Catholicism, and western religions. Basically the reasoning behind that is to put all your faith in god to take care of your urges, and to keep you pure, and keep you away from the devil.

That, the devil is behind every instinct, and makes us do the horrible acts we all do. This is all yet another sign of insanity. The Catholics who labeled our instincts as sins are about mild to medium insane. There is no god and no devil. .

Through evolution, these instincts are embedded into our D.N.A. and what we are born with. Through a few centuries, and civilization combined with the government has created laws designed to repress our instincts. What would happen if the laws, and consequences were lifted? Can humans be trusted without the consequences?

The current government and the justice system haven't worked in many centuries, if it was working, you wouldn't see the chaos that you see in everyday life. The police are followers to a corrupt system. The police do not protect us.

The police are just band aids to the legal system. The true security has to derive from what we project from our I.D.'s, making the right choices, in our mental thought processes. Civilians are part of the problem with this too, when humans don't have any consequences, and trusted with freedom, we abuse it, we act irresponsibly with the free will we are born with.

Society does not like how the government apprises us what we can, and cannot do. One current issue is, and has been for a long time is parenting. Parents apprise their children what to do, not the government. Dating laws are also very turbulent, the government does not obtain the right to invade or construct our desires.

The constitution may not be what you think it is, the constitution is an expression of your values, and beliefs, and your visions. It becomes a criterion for which you measure, and value your life. This constitution which was made, and signed by several of our forefathers, there are 7 articles which is called the articles of confederation, and the bill of rights which is the first 10 amendments.

Most of the constitution is meant to serve, and protect society, however, government officials have ratified much of it since. All agents of the government are supposed to serve us, or at set an example. That creates another controversial issue that may never be solved either, which are the values, and beliefs that one possesses about providing for oneself. Should there be one leader to rule a country? Should there be one system to rule the country?

One factor in leadership that Thomas Jefferson taught was the authoritive figure obtains that position to teach or serve the subordinate. Without the people under the one in charge, the one in charge would not hold that position. Most of the responsibility for setting the example falls onto the person with the authority, the one in charge. The rest of them, are responsible to go the person in charge when something arises. Maybe come to a solution together.

Most civilians are sheep, and follow the herd, which may be an answer in why America has been in a massive decline. What would happen if civilians stopped voting? If everyone stops voting, civilization would eventually crumble. Agents of the government depend on us to keep them in their position.

The government has become what Thomas Jefferson feared, the government has become legalized criminals. The government is operated by functioning psychopaths, they only empathize for themselves. Most of them are medium insane if you remember the insanity chapter earlier.

Censorship is another example of the intrusive behavior the government has displayed, and presented. Some parents blame the station, or the band, or the channel on TV, the parent decides what their child can see, not the government. No such thing as inappropriate, that's what called "socially acceptable" by the social norm.

Here are the many factors and virtues in what makes a real leader:

A real leader chooses to obtain this position to protect, teach, and serve the subordinates.

The leader requests feedback to improve their skills.

The leader gets to know all their subordinates, and the subordinates strengths, and weaknesses.

It is one of the leader's duties to provide assistance in the escalation of their subordinates strengths and weaknesses.

The leader holds the subordinate accountable for their actions or choices.

The leader will praise in public and punish in private.

If the leader observes potential for the subordinate to move up or develop more skills, the leader will send the subordinate elsewhere to do just that.

Mental discipline, your mental thought process, think positive thoughts when stressed. Stress is self-induced. Prioritize some tasks you will have to wait on. Remind yourself it's not the subordinates fault if you're stressed. Find a different outlet.

Leaders don't choose who they help. It's vice versa.

Teach a mind, don't yell at a mind.

Sometimes no matter the decision made, you will anger or let someone down. It's impossible to satisfy everyone, every time. Those are some traits of leadership.

We have to train our subconscious, and our subconscious. When we have made a mistake before, and we are in that situation again, our subconscious will recognize the image or sound, or the external stimuli that it is, and it will register in our brain, for us to make the conscious decision, not to make the same choice as last time. An accumulation of this will divert you away from a level of insanity.

Chapter 3

Humans are productions of the Universe Theory

This is one of the mind bending, ground breaking theories of Johnathan's I was telling you about earlier. No one else has thought about this that I know of, but Johnathan. He figured this out through his own studying, and writing on paper over the years. Over the years he has modified it, and added more substance to the theory. This is more to do with his quest on where humans came from, Johnathan doesn't believe in god, and the next 2 chapters go hand in hand. So here it is as follows, Johnathan's theory.

Humans are matter, and so are literal objects, and objects emit energy. Underneath this energy are atoms, underneath that are protons, neutrons, and electrons, under that is quarks, under quarks are the strings of energy.

Ever notice how people can't seem to sit still for a long period of time before it takes a mental effect or toll. You may have noticed how there's no such thing as bottling up emotions, emotions will manifest somehow. Humans are productions of the universe, along with our imagination, and how we can conjure up a thought.

Now currently there are others possibilities on how it all came to be, for that time being, the big bang theory still upholds #1. Before it all came to be, space became too dense, and tight, and space collapsed within itself, and exploding. God didn't put space here at all, space is space by itself. There are many similarities between humans, and science, and the universe.

Our bodies are the hired guns for our minds. Our minds cannot explode so our bodies lash out. Anything can happen through time, the formation of stars, galaxies, events here on earth, geography, you name it, is due to two factors, pressure and time, or entropy, which is an object or substance in perfect form until it comes out of it's perfect form, breaks, melts, freezes, etc. Anything happening or coming together is a phenomenon set in course from the big bang.

Another example is when people spend time together, people slowly growing mentally closer, and may not realize it. Matter is composed of molecules, which is what each person, each item, each object is composed of, and has its own gravitational pull, gravity is created by the atoms in matter, and it is an energy woven in the strings of space-time. Every piece of matter has atoms, inside atoms are protons, and neutrons, and electrons, and a nucleus. Just like the solar system.

The sun is the nucleus, the planets after that orbiting around the sun are the proton, and the neutron, and the electrons, (even though electrons do not orbit around the nucleus) and even deeper inside those, are quarks. Humans are productions of the universe. A proton is positive charge, and neutron is neutral, and electron is negative. Just like our minds, we have positive thoughts, and negative thoughts.

Each object in the whole universe has its own gravitational pull. The bigger the object, the stronger the gravitational pull, and the farther away, the weaker the pull is, on both humans and object. We just don't detect this is all. Our brains are not yet evolved enough to notice this.

With humans, the gravitational pull is mostly mental. More time we spent with each other in a closed environment, the closer the people will grow together mentally, and emotionally on a subconscious level. Gravitation is a consequence of space time. Space time is a mathematical model that combines the two into an interwoven continuum. It may take objects a few thousand years to draw closer but they eventually will.

Nothing in this universe is meant to stay in one place forever. Our imagination is just like the universe. Our imagination is called our minds eye, rooted in our brain stem. Our central nervous system, our 5 senses, anything we touch, hear, smell, taste, or see, registers in our subconscious to the appropriate part of our brain that is responsible for that function.

There is no god, no Jesus, no Mary, no heaven, no hell. Humans have positive instincts too. How millions believe in god, that he's all around us, the Holy Spirit is all around us, Johnathan has an answer to this, humans feel the god within themselves. Human's project god from the positive instincts felt inside our I.D.

These instincts called, tranquility, peacefulness, serenity, all good instincts we feel inside of our subconscious. They somehow slip through the Ego or the I.D. A negative monster in all of us, and there is the positive monster in all of us. Which will you be in the question you ask yourself? Man created god and Jesus in our image.

Quantum physics, well, at least some of it, is the study of the behavior of particles, and relationships, and how particles interact and everything, this is just like humans. Our environment is not what many think it is, there is a reason why our brain has many parts and functions. Homosapiens are not yet evolved enough to perceive the world as is, without the parts of our brains, and its functions.

Your eyes, hands, ears, and tongue, etc. are only the receptors. The occipital lobe sees it, the temporal lobe hears it, and other part of the brain smells it, the appropriate part of the brain feels it, organizes together the pieces of the external world. Certain parts of our brain make up or subconscious, and certain parts make up our conscious. When our subconscious organizes the external world, our conscious receives it how we do every day.

When you turn on the light switch in a room to see, light is a radiant energy. It is a form of electromagnetic radiation that is invisible to the human eye. By time you see the light is already traveled around the earth several times. Infrared are longer wavelengths, and ultraviolet are shorter. Speed of light is 186, 282,000 miles per second.

The behavior of electromagnetic radiation depends on its wavelength. Higher frequencies have shorter wavelengths, and lower frequencies have longer wavelengths. When electromagnetic radiation interacts with single atoms, and molecules its behavior depends on the amount of energy per quantum it carries.

Electromagnetic radiation in the visible light region of the spectrum, consists of photons that are at the lower end of energies that cause electronic extinction within molecules, which leads to bonding of the molecule, at the lower end of the spectrum. EMR (infrared) is invisible to the human eye, its photons no longer have each individual energy to cause a lasting molecular change in the visual molecule retinal in the human retina which triggers our sight. Depending on the behavior of the photons wavelength that hit objects or matter, and matter has atoms inside. Wavelengths have crest and troughs, the distance between the waves when the waves hit the object that pigments, determines the "color" in the color prism. In each object, there are pigments that circulate, and play connect the dots. Color does not exist in the external world, our brain organizes the light we see, as we see it.

Color is yet another example on how the external world does not exist how we see it to be. Our internal world (meaning our mind) organizes it for our species to see, and understand. So psychology and science can go hand in hand. Psychology is the science of the mind.

Explaining how the external world, and the internal world works, lets will dive into our senses. Sound is a vibration that propagates as a mechanical wave of pressure, and displacement. Mechanical wave is a wave that propagates as an oscillation of matter, and transfers energy through (here's the kicker) a medium. Not many people know this part.

What these mediums are, is a transmission medium or a material substance. Solids, liquids, and gasses. Transmission mediums for sounds are usually through solids, and liquids. Sound is reception of such waves, and their perception by our temporal lobe in our brain, located at the bottom of our brain.

Sound propagates through air, water, and solids as longitudinal waves, and is also as transverse waves in solids. Longitudinal sound waves are of alternating pressure deviations of rarefaction. Transverse waves are of shear stress at the right angle to the direction of propagation. Sounds waves that are plausible by humans has frequencies from 20 Hz to 20,000 Hz. The speed of sounds depends on the medium.

The sound source say comes from a stereo, this sound creates vibrations in the medium through say a couch, air or the coffee table. As the sound vibrates the medium, the vibrations propagate away from the source at the speed of sound, forming a wave. The behavior of sound propagation depends on the relationship between density and pressure. This relationship is affected by temperature, which determines speed of sound within the medium. Our ears are the receptors, the ear canal, equilibrium and eardrum, etc. Our temporal lobe is what is really hearing it.

The next senses to cover will only be comprised of the high lights on how we perceive it all. Such as touch, begins with receptors in the outer layer of the skin called the epidermis. Nerve endings here or just below the epidermis cells respond the various external stimuli. That's what all these senses are, external stimuli. These sensations go through a network that sends electrical signals to the spinal cord to the cerebral cortex in the brain.

Smell, the smell of a substance has to evaporate through the air or a vacuum, and travel into your nose, which is the receptor, and past your nose is a stump like area called the orila. Look like little hairs, and attached to the cilia. The signal goes to the appropriate part of the brain.

Taste, on our taste buds, receptors that open up to receive the molecule of food, once molecule hits. Through electrical networking, goes to the right part of the brain. So the appropriate part of the brain organizes the external stimuli into the internal world for us to see.

Chapter 4

Humans may be particles

This next chapter is how Johnathan came to the conclusion that humans are/possibly particles. Quantum physics is the study of the behavior of particles and their relationships. Our brains are not equipped to observe this, the invisible space if you will.

Ever wonder what the invisible space is? Or why we can't see the invisible space? Time very well may not exist, although we use it as a calculation every day. Life paths, time, quanta state, and the spiritual realm are all tied in together.

Humans have much in common with particles, particles can be become or are humans. Inside atoms are the center, which is the nucleus, then orbiting the nucleus are protons, and electrons, and neutrons. Protons are positive, electrons are negative, and neutrons are neutral. Now, there is something taking place here within the atom with the protons, electrons, and neutrons, they all orbit around the nucleus in what is called atomic orbital.

Now, atoms are in a circular shape. This is where geometry comes in, in circular objects such as atoms that have a radius, circumference, and diameter. Diameter is the length of the inside of the atom, circumference is the length on the outside of the atom, and radius is the length from the middle of the atom to the edge. Now normally a person would use these functions to solve an equation for a circular object, but there are rotations inside atoms. The electrons do not orbit the nucleus in the sense of a planet orbiting the sun, but instead exist as standing waves. The lowest possible energy an electron can take is therefore analogous to the fundamental frequency of a wave on a string.

Just like in our solar system, the sun is like the nucleus, and the planets orbiting the sun are like the protons, and neutrons. The angular momentum and atomic orbital is in our everyday life not just in atoms, angular momentum happens when we flush the toilet, and the water circulates in a clockwise or counter clockwise motion. The clocks on our walls in our homes are rooted to angular momentum, every rotation, every orbit, etc. is based on angular momentum.

Here's how, Brian Greene proposed a new theory in the science field many scientists are thinking that we live in a multi universe. The big bang still happened, just not in the way we think it did is all. The explosion didn't go in just one direction like we think it did. It also went the opposite direction too.

This new theory of Brian Greene looks to be pretty accurate, when the big bang went in both directions, this created multi universes, and very well may be true. Multi universes theory comes from string theory according to Brian Greene. Each universe may be as close as you are to the object next to you, or may be very far away as well.

Brian Green says within string theory, the strings that we are talking about are not the only entities that this theory allows, it also allows objects that look like membranes, which are two dimensional surfaces. Within string theory, is that we may be living on one of those gigantic surfaces, and there can be other surfaces floating out there in space. If we are living on one of these giant membranes, then the following can happen, when you slam particles together some debris from those collisions can be ejected off of our membrane and be ejected into the greater cosmos in which our membrane floats, he says. If that happens, that debris will take away some energy. So if we measure the amount of energy just before the protons collide and compare it with the amount of energy just after they collide, if there's a little less after, and it's less in just the right way, it would indicate that some had flown off, indicating that this membrane picture is correct. Now some other scientists say, this is like saying how the multi universes came to be.

Now along with multi universes, there is talk that we are living in the future or the past, and that alternate universes are in the opposite timeline. If we are the future, they are the past, and we can co-exist in the same universe or in separate, there is still speculation on that particular part. Now, going in to this deeper, there is a question on whether or not time exists, if it does exist, maybe these alternate universes are a more advanced breed of humans. Like mentioned earlier, our minds are not capable of seeing the invisible universe before our very eyes, so what if in some of these universes, the far future, the human brain can now see everything, and does not require sleep. Maybe there is an alternate universe in front of our eyes now, and the particles are the humans just in an alternate universe, and humans can see everything and use all of our brain power.

Think of time as a frozen ice rink, and we humans create time with our own life paths. The spiritual dimension and the invisible universe we know as quantum physics are the same universe, or perhaps they co-exist, perhaps that is where our entities travel to by the strings of space.

This talk that our physical universe is actually from the future, and there are many other universes we cannot see with our eyes. One way how the spiritual realm exists together with the quanta world is this: mediums, mean a human that has certain abilities to communicate with the dead, such as Johnathan Lee

There a few types of mediums, a trance medium, a physical medium, and a mental medium. A mental medium has the ability to hear, and feel entities of the spiritual realm, aloud, and in their own mind, and to communicate with their spirit guide, and receives messages, and visions in their heads telepathically, and communicates the message to the precipitant.

Now follow on this one, as discussed earlier, mediums with sound waves, and how the sound waves travel through the object or liquid or air known as the medium. These humans who obtain these abilities as a mental, trance or physical medium are the mediums for the spiritual world too. It's all the same concept, the human medium with the abilities, is the medium, receiving messages from the spirit guide, ghost, or spirit, and on to the recipient.

Here's the similar concept explained, the stereo is the entity, the object such as the coffee table is the medium, just like the human medium is in the middle of the spirit guide, and the precipitant, and the waves are the messages going through the object, and you are the recipient just like in our world too. Time may be an illusion, and may not exist, our life paths are what seem to create time, as it seems to fly by. Our universe very well may be in the future.

Chapter 5

A State of Mind/Sanity

This is where Johnathan has experimented upon himself by staying up insane hours, (pun intended) and feeling the effects of it. How he functions better with little to no sleep, and how he discovered that people with the highest of i.q.'s have the ability to function better with little to no sleep, but in order to do this, they have to be able to discipline themselves. Johnathan has pushed his mental capacity to the limit, he had to be sedated in order to fall asleep.

Ever wonder what lurks in your subconscious? You find out every time you fall asleep and dream, (if you do dream that is). This is our instincts, and our external stimuli. Throughout our conscious waking hours, our brain has alpha waves generated in the thalamus part of our brain. These waves are slowest in our waking hours. Our external stimuli are absorbed by our subconscious from the external environment recently talked about during our waking hours.

During our waking hours our instincts emerge. Impulses like anger, anger is a response to a threat from external stimuli.   We feel these impulses and instincts every day. Some of us are environmentally conditioned not to act out on these instincts. Therefore, we consciously make the effort to repress or suppress that emotion, and let it out somewhere else.

So that's a factor, we dream to suppress. Another factor is repression, like mentioned earlier, our minds are not evolved enough to handle the external environment during our waking hours. As we sleep our body relaxes and so does the mind.

The delta waves emerge, and go through the roof when asleep. Some people question why. Here is the answer, because our subconscious is coming to the light. Like a volcano if you will. You see the dream in your mind's eye, why? Of all the external stimuli you absorbed during the day.

The sleep stages: during the earliest stages of sleep you are still relatively awake. Brain produces delta waves, which are small and fast. When the brain relaxes, slower waves known as alpha are produced. During this stage you will feel like you are falling asleep or hearing voices. Now, supposedly these are hypogyny hallucinations. No, you're not hallucinating, during this state of mind one is in, you're not entirely awake, nor are you entirely asleep.

These are not hallucinations, no such thing, it's the spiritual side. Stage 1, is the light stage. For light sleepers, they wake up to every external stimuli, noise, mostly. This stage is in between wake, and sleep. The brain produces high amplitude theta waves, which are very slow.

Stage 2 lasts 20 minutes, and the brain produces rapid rhythm wave activity, sleep spindles. Body temperature decreases, and begins to slow. Light sleepers have very few sleep spindles. Stage 3, deep slow brain waves called delta waves emerge. Now, you are unresponsive to external stimuli and it fails to generate a response. Sleep walking sometimes occurs in this stage as well.

Stage 4, is when dreaming emerges or REM rapid eye movement. Increased respiration rate, and brain activity. Like mentioned earlier, this is when your subconscious is emerging, and you see it in your mind's eye, of all the external stimuli you have absorbed or repressed internal emotions. Sleep doesn't happen in order. The process unfolds like so, 1, 2, and then 3. After 3, 2 is repeated then 4 for REM. Then after 4, goes to 2. REM lasts about 90 minutes.

Sleep deprivation, something we have all experienced from time to time. No such thing, it's a state of mind, when we become sane. We can see, and hear more, the external world, our minds use more brain power in this state, and we are leaving insanity the longer we are awake. The waves in our brain that emerge when we sleep are the same when we have been awake for a long time, and we can use that brain power. When awake for so long, you're not asleep, but you're not awake either. You're in that "twilight" stage of sleep discussed earlier.

A person is in that in between stage before stage 1 officially happens. That's when some people feel cold, or experience the "buzzing" sensation feeling. One feels the same "buzzed" feeling when under the influence of alcohol, both of which are altered states of consciousness or state of mind. Believe it or not, some people find it easier to remember, focus, and function better on a severe lack of sleep, possibly due to increased brain activity

When in the altered state of mind/sane, there are stages to this, becomes worse by the day, nothing significant happens until between 2 to 3 days. At this point, paranoia may emerge, colors may start blending in with each other, and may start to hear indistinct voices, and images. None of this is hallucinations, you're in an altered state of mind/sane, which is starting to hear, and see the external world, and use more brain power.

People don't need drugs to see or hear the external world or use more brain power, severe lack of sleep will do that just fine. Careful with this method though, it may lead to a "sleeping disorder", after so long of going days without sleep, this will break your circadian rhythm, and body clock, and develop a "sleeping disorder". Again, no such thing as sleeping disorders, just going by language we all use.

At 72 hours of no sleep, you are declared legally insane. When we go "insane" we are actually going "sane" we are hearing, and seeing the external world, what others cannot see. Insanity, and sanity are a state of mind. It's vice versa how we are told it is.

Now this figure isn't 100 % of the population, but more people who aren't in the altered state of mind are the insane, and more people who are in the altered state of mind are sane. Doesn't last long though, we all need sleep eventually, and our organs depend on it, more so from the physiological reasons, our organs are controlled by our brains.

It was Sigmund Freud who came up with the two parts to the dream.   Which are the latent, and the manifestation. The latent is the part of your dream that is so disturbing your Superego or your Ego decides to block it out, meaning, the tunnel to your conscious or your actual conscious decides not to let it to your awake state. The manifestation is the part of your dream that you remember when you first wake up.

The only way to recall the latent part of your dream, is when you are awake, certain images, sounds, smells, touches, etc., will trigger an image, sound, smell, touches, etc. from the latent. It's in all in our senses, neuroscience, and in the mind. Now, let's bring the paranormal into this.

Ghosts, spirits, demons, spirit guides all exist. That realm does exist within a quanta state. This realm is not what you think it is from the movies, and TV shows. Some elements are true some are not from the shows, and movies. It is very possible for ghosts, and spirits, and demons to exist without god, heave, hell, and angels.

You the reader have had experiences where you heard voices in your dreams or images of a deceased loved one and such, these are not hallucinations. These are very real, the deceased loved one is simply reaching to you in your dream. The message may be an image of the person, or voice, etc. When you're in that "twilight" sleep before the actual sleep starts, you are most vulnerable to the paranormal.

Why? Because your subconscious is slowly getting ready to come to the light. Our brains are not yet equipped enough to see the paranormal or hear them, (unless you're a medium or psychic), in that "twilight". Your subconscious brain waves escalate, and your true brain abilities emerge, this has nothing to do with god or Jesus.

Chapter 6

What the Bible doesn't Tell You, and the Paranormal

Now this chapter in particular Johnathan has never liked the bible in any way. It's fascinating to read about it, but at the same time he strongly believes none of this happened. This chapter is for those who do believe in god, and do believe in the bible. Like mentioned earlier, Johnathan formed his own belief how the paranormal exists without god existing. In this chapter, he is basically giving believers of god the finger, and asking them how can you be so blind, and believes everything without question? It is as follows.

The legend goes, god created archangels, 4 of them, Michael who is the oldest, Lucifer, Uriel then Gabriel. Then god created humanity, and wanted the archangels to bow down to humanity, when humanity was clearly seen to be flawed. Lucifer went to god, and refused to bow down to humanity while Michael and the other archangels did.

God then casted Lucifer out of heaven to earth, and on earth, Lucifer discovered a human named Lilith. People may not be aware that Adam had a first wife, before Eve. Lilith was his first wife, it was Adam, and Lilith at first. The couple fought over sex positions along with other arguments about dominance. Adam wanted to be the dominant one, and Lilith wanted to be the dominant one as well.

Lilith ended up leaving Adam, and god sent 3 of the archangels after her to retrieve her, but Lilith refused. Lucifer discovered Lilith, and she is already vulnerable. Lucifer manipulated her to come with him. She did, and Lucifer created hell in spite of god. God did not create hell, Lucifer did, in rebellion. And Lilith through and twist became the first demon.

Hell became the place where demons were produced from twisted and tortured souls. On earth, god took Adams rib, and made eve. Adam and Eve had at least 3 sons, Cain, Abel, and Seth. When god asked Cain, and Abel to sacrifice an animal, god was more pleased with Abeles sacrifice, then Cain's.

Cain grew jealous of Abel, and felt angry as well. Cain took Abel to a field, and killed him with a weapon. Some say Lucifer made him commit the terrible act. Which it was not Lucifer, the act was done all by Cain, and his anger. God then cursed Cain with what is called "the mark of Cain". The mark is to warn others that killing Cain would provoke god's vengeance or if harm was done to Cain, damage would come back to the harmer.

The source of the power is the blade Cain used to kill Abel with. According to the legend, Cain killed Abel due to Cain feeling Lucifer was corrupting Abel, and if Cain felt if he kills Abel, then Abel's soul would go to heaven, and not hell to be tormented. Back to Lilith, when she fled from Adam, and went to the red sea where the archangels found her, she was told her children would be doomed to die. Lilith said she would take the lives of Adams children, and she did in a way. Lilith did go back to Adam, and their 3 children were all demonic, due to the fact that Lilith was a demon.

No such thing as heaven, or hell, or at least not how we are told that it is. When a person dies, and the person is at peace with their life, and the person knows they have died, the deceased becomes a spirit, and spirits cross over to the quanta universe/state. A spirit is the energy manifestation of the energy inside of our brain, D.N.A., and possible our positive instincts, that manifests when we die. A spirit is tranquil, peaceful, and serene. The good instincts have taken over 100%, the complete opposite of a demon.

Ghosts, these entities are in between, these entities are not at acceptance for variable reasons. They are angry, they do not desire to be dead, and they don't know they are dead. This sounds just like a human, doesn't it? Humans can be dead inside too, emotionally that is.

A demon, is not from hell. The hell is inside of the ghost, the anger, the misery, just like what humans feel, and if a ghost resides in that stage long enough, its humanity will be stripped away, and become a demon, left with nothing but the bad instincts. The paranormal universe does exist, the entities in that universe may see us, and how we see them. In that universe exists spirits, ghosts, and demons.

There are no angels, but spirit guides. Spirit guides do not protect us from physical harm, but guides us through our life path. Spirit guides are most active in our life when a major life change is about to happen. The bible says that mediums are bad, for variable reasons, however, that is only due to god wants humans to desire nothing else except him, and among other reasons. Mediums communicate with their spirit guide or with the dead. The result can be dangerous yes, if not communicated correctly.

A shaman is someone who when in a state of mind is vulnerable to the paranormal. When talking to the paranormal, whatever entity is at that door to come to our world, can be the dangerous part. That's where the medium or shaman communicates to their spirit guide mentally or aloud to protect him/her and the people around him/her. Good prevails over bad every time.

Chapter 7

Society is the Asylum

This chapter is yet another chapter in which displays Johnathan's mediumship abilities, and how he observes so much in society that very few others do. Johnathan feels like Dr. Seuss's "How the Grinch stole Christmas". The Grinch was clearly much smarter than the Who's down in Whoville, and he was clearly "different" to the rest of them. How the Grinch chose to live away from the society in a mountain, isolated. Now, the whole story of the Grinch is different than Johnathan's of course, but those are the similarities.

Now, Johnathan is not grumpy like the Grinch was, oh but the opposite. However, Johnathan feels social phobia in public, and his social skills have declined through time. He is a friendly person, but what goes on his mind, and communicating that, is his main problem that he struggles with. Johnathan does not like what society has declined to or the conformity that comes with it, and he feels people make excuses for their behavior

Johnathan feels again this is where humans are delusional, blaming their own actions, decisions, and behaviors on what was "expected" of them back then. When in reality, no one has an answer on why society was so uptight many decades ago, but Johnathan will give you an answer to that, and so much more, as follows.

The human race is mostly sheep, and the government or the so called "higher powers" are the shepherds. Television, commercials, magazines, media, marketing, sales, advertising, radio, are all tools to lure us to succumb to the material of the society. Newspaper came first, then radio in the 1920s. NBC was the first station to be on the radio, then ABC, then CBS.

Television came out in the 40s or 50s, depending on where you lived, and your income. There were only at most 3 channels back then, and advertising had been around longer than the 1900s. Just the public airwaves made it easier to reach a higher volume of an audience.

Television is just a brainwashing tool. Commercials especially, ads for material we don't need to impress people or ourselves that we don't like. The material possessions we own, end up owning us. It's all in the marketing, ads, and presentation is the biggest one. Presentation is how they lure you in, through manipulation. They are the deceivers, and we the consumers are the believers. What would happen if we all stopped buying material?

Consumers are in charge, not the producers, its supply and demand. If consumers stopped demanding it, the producers can't supply it.   They money we spend every day on material, gets to the government eventually via taxes, and the government has not spent our earned money wisely.

Society serves the government through our everyday life. They are legalized criminals, we are living in a gigantic asylum, every building you see, every business you see, a branch of the government has to approve of that building to make money. Much of the human race has displayed humor in other people's misery.

The people in presentation, marketing, and advertising are functioning psychopaths, they just present themselves not to look that way. Business, in Johnathan's opinion is sociopathic, and presents an absence of logic. Business has presented nothing, but a covet for money, and the absence of empathy, this is backwards to how society should be.

Every business that is opened, the money made there goes back to the government via taxes. Every commercial you see about material such as hygiene, or hardware, or new cars, electronics, dating websites, clothes, you name it, everything that has to do with purchasing material goes to the government eventually, via taxes.

Johnathan feels very strongly against marijuana, perhaps for personal reasons, he has seen former friends of his ruin their lives from weed. However, marijuana is a very controversial subject in America. Supposedly it has been filtered for medicinal reasons, Johnathan is skeptical about this, he desires proof of the filter for marijuana. Johnathan says if someone thinks it should be legalized, that person should go smoke it. (In a challenging tone)

Houses can exist without money, neighborhoods can exist without money, and cars can exist without money. Money is not a requirement to make the houses, cars, or any objects. Money is not required to make objects, an individual can literally, physically, make a house without money. Takes an awful lot of construction, etc. but from a physical stance, it can be done without money.

Morally, and instinctively this society has greatly declined in many aspects. Social conventions are another factor, partying at clubs, and bars, etc., the loud music, and the alcohol, and the lighting or electromagnetic radiation are all manipulation tactics used to manipulate you, to spend your money. Partying is fun, but is it worth it?

Each tax form you fill out for you job, there's a reason for it and it's a sociopathic trap. Each question is a trap, asked for a reason, your gender, you race, ethnicity, if you're married, if you have children, if you own any land, or bonds, stock, 401k, etc. We all have been asked these questions before. Here's the answer for it, if you're a young single white man or woman, the government doesn't care about you, why? They don't get their tax credit from an employer that comes from the employee or the beneficiary.

Marriage is both legal and religious. This is a legal bind more than anything else. The government shouldn't say whether or not a couple will be together forever. The child born without the parent's in marriage is a very religious belief. It violates biblical laws, many western religions.

OSHA is the government, and the harassment laws are backwards. It is only harassment if the colored races are the ones to be harassed due to the employer, and the government covets their tax credit. College is another example of the trap the government set for society. Though commercialism, and in these commercials, we see programs for college, and people in the commercials encouraging you to go back. That commercial had to be produced, and made somehow, through money, and through advertising. By law, there certain requirements to go through to get the commercial approved.

Most people take out student loans, and student loans ultimately come from the government. Weather a person receives it from a bank or the actual government itself through your state workforce website. Then you have to pay back that student loan, and even those who do obtain their degrees, they will live the life that has been expected. The reality is, a person doesn't need a degree to be an expert in a subject, but that's how society views it, in black, and white.

Censorship is another variable in the equation that needs to be resolved. First off, no such thing as inappropriate, those who believe in this, are sheep. Inappropriate is just a term from the social norm of society for socially acceptable. Inappropriate is usually about sexual nature. Christianity, and Catholics decided the social norm, and not to allow those words without censorship centuries ago. Much of societies back round originally come from religion.

Chapter 8

Holidays are Not What They Seem

Now this chapter, Johnathan feels passionate about. It relates to the rest of the book, but in a subtle hidden way more so, and relates to the previous chapter. He feels that society doesn't understand what these holidays are, the origins of them, or simply the holiday is not what it seems. He feels that society has been brainwashed over however many years through commercialism, and manipulation, and among other tactics.

Johnathan feels that since he refuses to believe in holidays, he is labeled. Now, he still celebrates them, but not for the reason we are all told to though, Johnathan like anyone else wants to have fun, so that's what he does during these holidays. He feels that those people who do believe in holidays are a sign of insanity. He also feels that humans shouldn't need a holiday to express their appreciation or gratitude to each other, or for family to come together.

He feels that much of society's behavior is an image, and people only behave in a positive performance during the holidays. The avarice never ends in society, and people have been lured in by commercialism, and greed. Johnathan feels that people should present empathy for each other, and be possessed by greed.

New Year's Eve/ day is the first one to talk about, the first day of the year on the modern Gregorian calendar, as well as the Julian calendar. As a date in the Gregorian calendar of Christendom, New Year's Day liturgically marked the Feast of the Circumcision of Christ, and is still observed as such in the Anglican Church, and Lutheran Church. New Year's Day was founded by Pagan traditions. New Year's was founded in the 7th century by the Flanders, and

the Netherlands of the Pagans. It was a custom tradition to exchange gifts as part of the New

Year's.

This holiday is not much like the others to be discussed, as this holiday is simply

celebrating another year of life. By celebrating we are alive, and we survived another year of

life. Next holiday is up is V-day, and no, not the s.t.d., Valentine's Day. This holiday is based off

a saint.

St. Valentine's Day began as a liturgical celebration one or more early Christian saints

named Valentinus. Several martyrdom stories were invented for the various Valentines that

belonged to February 14, and added to later martyrologies. A popular hagiographical account of

Saint Valentine of Rome states that he was imprisoned for performing weddings for soldiers

who were forbidden to marry, and for ministering to Christians, who were persecuted under

the Roman Empire. According to legend, during his imprisonment, he healed the daughter of

his jailer, Asterius. To this story states that before his execution, he wrote her a letter signed

"Your Valentine" as a farewell. Thus you have the whole "would you be my valentine"

tradition.

Next up is St Patrick's Day, and again, this is not a major holiday, but still celebrated in

our society. Saint Patrick's Day was made an official Christian feast day in the early seventeenth

century, and is observed by the Catholic Church, the Anglican Communion especially the Church

of Ireland, the Eastern Orthodox Church, and Lutheran Church. The day commemorates Saint

Patrick, and the arrival of Christianity in Ireland, as well as celebrating the heritage, and culture of the Irish in general. Celebrations generally involve public parades and festivals, céilithe, and the wearing of green attire or shamrocks. Christians also attend church services, and the Lenten restrictions on eating, and drinking alcohol are lifted for the day, which has encouraged and propagated the holiday's tradition of alcohol consumption.

So the modern cultural social convention of the indulgence of alcohol, was evidently back in that time too. Much of what is known about St Patrick comes from what written by Patrick himself. It is believed that he was born in Roman Britain, in the fourth century, into a wealthy Romano-British family. His father was a deacon, and his grandfather was a priest in the Christian church. According to the Declaration, at the age of sixteen, he was kidnapped by Irish raiders, and taken as a slave to Gaelic Ireland

It is said that he spent six years there working as a shepherd, and that during this time he "found god". The Declaration says that, god told Patrick to flee to the coast, where a ship would be waiting to take him home. After making his way home, Patrick went on to become a priest. According to legend, Saint Patrick used the three-leaved shamrock to explain the Holy Trinity to Irish pagans.

According to tradition, Patrick returned to Ireland to convert the pagan Irish to Christianity. The Declaration says that he spent many years evangelizing in the northern half of Ireland, and converted thousands. Tradition says that he died on 17 March, and was buried at Downpatrick. Over the following centuries, many legends grew up around Patrick, and he became Ireland's foremost saint. On St. Patrick's Day, it is customary to wear shamrocks and/or green clothing. St Patrick is said to have used the shamrock, a three-leaved plant, to explain the Holy Trinity to the Pagan Irish.

Easter is next, the major holiday. Now, there are a few ways to look at Easter, if you're a believer in god, then it's a whole different ball game. Then there's the Easter bunny, and how that came into the story, and for the non-believers. If you don't know already, most holidays, are from religious back grounds. Easter is celebrated as the resurrection of Christ after he was crucified. So this has a brutal history behind it. For the non-believers of Christ, Easter is not a holiday, but non-believers of Christ still celebrate anyway.

The Easter Bunny is a folkloric figure, and symbol of Easter, depicted as a rabbit bringing Easter eggs. Originating among German Lutherans, originally played the role of a judge, evaluating whether children were good or disobedient in behavior at the start of the season of Eastertide. The Easter Bunny is sometimes depicted with clothes. In legend, the creature carries colored eggs in his basket, candy, and sometimes also toys to the homes of children, and as such shows similarities to Santa Claus or the Christ kind, as they both bring gifts to children on the night before their respective holidays.

The hare was a popular motif in medieval church art. In ancient times, it was widely believed as by Pliny, Plutarch, Philostratus, and Aelia that the bunny was a hermaphrodite. The idea that a hare could reproduce without loss of virginity, led to an association with the Virgin Mary. This holiday is very similar to the concept of Santa clause.

Next up is 4th of July, this holiday most people should know by now, is the declaration of independence. Our founding fathers, back in 1776, signed us, America, officially free from other countries. This is a celebration of freedom though America.

Next up is Halloween, this becomes rather deep, bur surprisingly not the most disturbing of the holidays. This day is remembrance of the dead. According to many scholars, All Hallows' Eve is a Christianized feast initially influenced by Celtic harvest festivals, with possible pagan roots, particularly the Gaelic Samhain. Other scholars maintain that it originated independently of Samhain and has solely Christian roots.

Samhain/Calan Gaeaf marked the end of the harvest season, and beginning of winter or the 'darker half' of the year. The souls of the dead were also said to revisit their homes. Places were set at the dinner table or by the fire to welcome them. The belief that the souls of the dead return home on one night or day of the year seems to have ancient origins, and is found in many cultures. In 19th century Ireland, candles would be lit and prayers formally offered for the souls of the dead. After this the eating, drinking, and games would begin. Throughout the Gaelic, and Welsh regions, the household festivities included rituals, and games intended to divine one's future, especially regarding death, and marriage.

The traditional illumination for geysers or pranksters abroad on the night in some places was provided by turnips or mangle worsens, hollowed out to act as lanterns, and often carved with grotesque faces to represent spirits or goblins. In folk lore, the jack o lantern, is a symbol that represents a soul, that has been denied of entry to both heaven, and hell. All of this is of believers of god and religion.

Today's Halloween customs are also thought to have been influenced by Christian dogma, and practices derived from it. Since the time of the primitive Church, major feasts in the Christian Church such as Christmas, Easter, and Pentecost had vigils which began the night before, as did the feast of All Hallows. These three days are collectively referred to as All Hallowtide, and are a time for honoring the saints and praying for the recently departed souls who have yet to reach Heaven. So of course, children don't know the real reason behind Halloween, the creepiness of it, and through history, society or Christians have covered up certain customs.

Thanksgiving is up next, now, Thanksgiving is the celebration of 1621 of the Pilgrims, and the Puritans in Massachusetts. The holiday has been celebrated as the 4th Thursday of November by a presidential proclamation back in 1863. There is speculation, and debate on whether this is the true origin of thanksgiving. Theories of that go into slavery, and brutality, although have not been proven. Giving thanks, has always been a custom for Thanksgiving, although Johnathan feels that people shouldn't need a holiday to be thankful or come together.

Now, finally the most disturbing of all holidays, Christmas. Those who believe in Christ, Christmas is simple, the celebration of the birth of Christ. Now, even this is skeptical, there is a certain group of people back then who wrote the bible that left certain parts out, and added certain parts to cover up what really happened. Some say the birth of Christ was not in Decembe,r but in the spring. For believers Christmas is simple, hence the word itself "Christ" and "mass".

For non-believers, this is what really happened, the romans in the year 1, this goes to the classic all religions think their religion is the right one, and hate all others. At some point in time, the Romans and another group called the Saturnalia crossed paths in civilization and Pagans had long worshipped trees in the forest hence the Christmas trees, or brought them into their homes, and decorated them, and this observance was adopted, and painted with a Christian veneer by the Church.

The Mistletoe Norse mythology recounts how the god Balder was killed using a mistletoe arrow by his rival god Hoder while fighting for the female Nana. Druid rituals use mistletoe to poison their human sacrificial victim. The Christian custom of kissing under the mistletoe is a later synthesis of the sexual license of Saturnalia with the Druidic sacrificial cult. The Christmas Present In pre-Christian Rome, the emperors compelled their most despised citizens to bring offerings, and gifts during the Saturnalia, and Kalends. Later, this ritual expanded to include gift-giving among the general populace. The Catholic Church gave this custom a Christian flavor by re-rooting it in the supposed gift-giving of Saint Nicholas.

The Origin of Santa Claus Nicholas was born in Parara, Turkey in 270 CE and later became Bishop of Myra, he died in 345 CE on December 6th. He was only named a saint in the 19th century. Nicholas was among the most senior bishops who convened the Council of Nicaea in 325 CE, and created the New Testament. The text they produced portrayed Jews as "the children of the devil" who sentenced Jesus to death. In 1087, a group of sailors who idolized Nicholas moved his bones from Turkey to a sanctuary in Bari, Italy. There Nicholas supplanted a female boon-giving deity called The Grandmother, or Pasqual Epiphania, who used to fill the children's stockings with her gifts.

The Grandmother was ousted from her shrine at Bari, which became the center of the Nicholas cult. Members of this group gave each other gifts during a pageant they conducted annually on the anniversary of Nicholas' death, December 6. The Nicholas cult spread north until it was adopted by German and Celtic pagans. These groups worshipped a pantheon led by Woden their chief god, and the father of Thor, Balder, and Tiw. Woden had a long, white beard, and rode a horse through the heavens one evening each autumn. When Nicholas merged with Woden, he shed his Mediterranean appearance, grew a beard, mounted a flying horse, rescheduled his flight for December, and donned heavy winter clothing.

Christmas celebrates the birth of the Christian god who came to rescue mankind from the curse of the Torah. It is a 24-hour declaration that Judaism is no longer valid. There is no Christian church with a tradition that Jesus was really born on December 25th. December 25, is a day on which Jews have been shamed, tortured, and murdered. Many of the most popular Christmas customs including Christmas trees, mistletoe, Christmas presents, and Santa Claus are modern incarnations of the most depraved Pagan rituals ever practiced on earth.

Chapter 9

Is this really the end?

Now this final chapter is all Johnathan speaking. I have done my job narrating this book for you. I would like to think I did a bloody brilliant job at narrating this book, as it is comprised of Johnathan Lee's theories, beliefs and his feelings that formed from his life path. Enjoy the rest my friends.

I cannot speak for others, but the best feeling in the world for me is when I make a positive impact on someone by doing a good deed to them. They are so overwhelmed they cry tears of joy, and they struggle to say "thank you" because they are crying. I have been on both ends of this, the giver, and the receiver. It feels amazing on both ends.

I love performing my mediumship for others, I get along better with the dead than the living sometimes. I do not fear the paranormal, for I have my spirit guide Emma on my side, and the positive always triumphs. I love performing a séance, which is me the medium, a witness or assistant that will assist me possibly by writing down on paper what I receive telepathically, and the recipitant, and this is usually performed at a table. If an individual feels there is an entity in their home, I covet to perform a séance, and detect whether or not there is an entity. As long as the individual who invites me into their home has an open mind, I will be at full capacity with my abilities.

Life is a journey, I believe it is both destiny and the individual's responsibility to make an effort that comprises a life path. Life is meant to be enjoyed my friends.

www.ingramcontent.com/pod-product-compliance
Lightning Source LLC
Chambersburg PA
CBHW080609180526
45168CB00007B/2842